Context Clues

5–6

Written by
Delana S. Heidrich and Stanley H. Heidrich

Editor: Pam VanBlaricum
Illustrator: Corbin Hillam
Production: Acorn Studio Books
Cover Designer: Barbara Peterson
Art Director: Moonhee Pak
Project Director: Linda Schwartz

Table of Contents

Introduction

Each book in the *Power Practice* series contains dozens of ready-to-use activity pages to provide students with skill practice. Use the fun activities to supplement and enhance what you are already teaching in your classroom. Give an activity page to students as independent class work, or send the pages home as homework to reinforce skills taught in class. An answer key is provided for quick reference.

Current research indicates reading for meaning is a multi-faceted task. Literacy requires mastery of a complex set of skills, including those involved in fluency, comprehension, vocabulary, context clues, inference, summary, punctuation clues, sequencing, figures of speech, story analysis, detail recall, and drawing conclusions. Direct instruction in each of these areas provides emerging and proficient readers the practice they need to develop and sharpen literacy.

Context Clues 5–6 is designed to provide students with practice in the art of locating and interpreting context clues. A clear understanding of context clues assists students in defining vocabulary, comprehending literal meaning, and deciphering abstract tones, themes, and concepts. The first seven activities in *Context Clues 5–6* provide direct instruction and practice in six separate context clue skill areas. The remainder of the activities combine the six skill areas to provide mixed practice. Mixed practice activites are arranged around related vocabulary from a variety of cross-curricular content areas.

The vocabulary and activities presented in *Context Clues 5–6* are based on best practices suggested by current research. The NCTE Standards for English Language Arts, Reading First, and the provisions of No Child Left Behind were all consulted in the development of these exercises. The selection of grade appropriate vocabulary was guided both by national grade level lists and the work of experts in the areas of context clues and second-tier vocabulary word selection.

Use all of the activities in this book or assign selected ones that fit into your current reading strategies curriculum. Assign them in class or as homework. Allow students to work independently or with a partner. However you use *Context Clues 5–6*, your students will be one step closer to success on standardized tests and mastery of an essential skill in the development and sharpening of literacy. Use these ready-to-go activities to "recharge" skill review and give students the power to succeed.

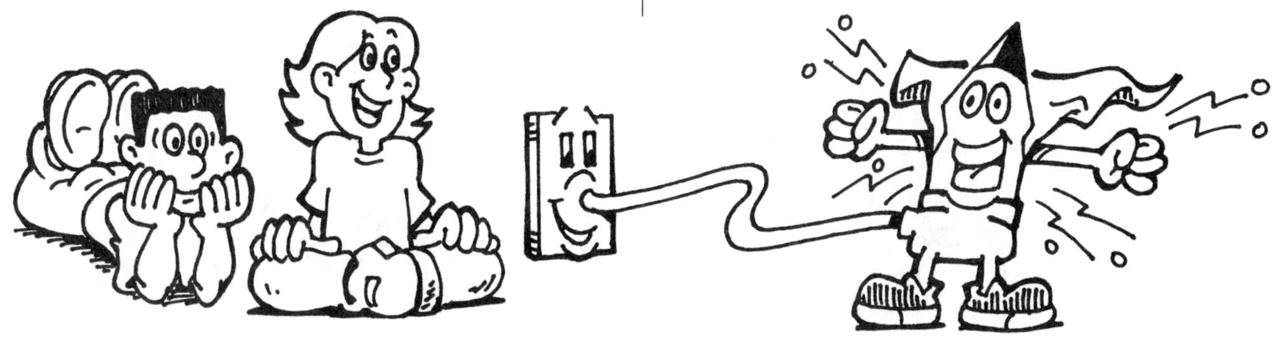

Name ______________________________ Date ______________

Word Part Clues

Knowing the individual parts of a word sometimes helps a reader identify the meaning of the entire word. In the italicized words in the sentences that follow, a word part that you should be familiar with is underlined. Use it to help you define the entire word. (Some root words drop a letter when a suffix is added.)

1. We got *adjoining* rooms at the hotel.
   - **A.** neighboring
   - **B.** spacious
   - **C.** cheap
   - **D.** clean

2. Selling popcorn at the ballgame might be *profitable*.
   - **A.** wise
   - **B.** unwise
   - **C.** able to make money
   - **D.** easy

3. Carrie found the hot sun *intolerable*.
   - **A.** pleasant
   - **B.** unbearable
   - **C.** bearable
   - **D.** bright

4. Tom is an *observant* driver.
   - **A.** poor
   - **B.** excellent
   - **C.** slow
   - **D.** attentive

5. Kevin actually enjoyed the *deafening* sound of the band.
   - **A.** great
   - **B.** extremely loud
   - **C.** strange
   - **D.** quiet

6. "Mal" means "bad," so I knew it wasn't good when Dad said the car *malfunctioned*.
   - **A.** ran out of gas
   - **B.** failed to work properly
   - **C.** moved slowly
   - **D.** got dirty

7. He works in the *industrial* section of town.
   - **A.** referring to selling
   - **B.** referring to making of goods
   - **C.** residential
   - **D.** older

8. Sterling was a *towering* man.
   - **A.** tall
   - **B.** short
   - **C.** stout
   - **D.** thin

9. *Disregard* that last comment.
   - **A.** remember
   - **B.** think about
   - **C.** ignore
   - **D.** study

10. Cathy's answer was *inaccurate*.
    - **A.** correct
    - **B.** incorrect
    - **C.** estimated
    - **D.** well researched

Name ______________________________ Date ______________

Example Clues

Sometimes sentences include examples that help a reader define an unknown word. Use the examples in the sentences below to help you fill in the blanks with the correct words from the word box. Look for these words and phrases that often indicate an example will follow: **such as**, **consists of**, **including**, **for example**.

WORD BOX

primitive
alternative
collaborative
audition
endurance
opposition
diverse
ambitious
controversial
vendors
grievances
feats

1. ______________________ energy sources include wind, solar, and water power.

2. The decision was ______________________ because it consisted of points that might anger people on either side of the issue.

3. ______________________ peoples such as the native populations in Australia and Africa have never used electricity or gas-powered vehicles.

4. Tim's ______________________ efforts, including adding illustrations and completing extra research, earned him an A on the assignment.

5. After his extremely disappointing day at the fair, Steven had many ______________________ including his complaints against long lines, misleading signs, and poor traffic control.

6. ______________________ at the amusement park sold everything including food, souvenirs, and necessities.

7. Susan and Megan's ______________________ efforts included researching together, studying class notes together, and writing the final paper together.

8. Your ______________________ for the role of Mary will consist of a scene reading, a solo number, and an interview with the director.

9. The school's ______________________ offerings include four foreign languages, three art classes, six literature offerings, and a class in horticulture.

10. Many of Superman's ______________________, for example, lifting trucks and leaping over tall buildings, are not humanly possible.

11. ______________________ to the proposal to cut down trees in the park is being heard from such groups as environmentalists, park goers, and tree lovers.

12. Kate saw several displays of ______________________ on the busiest shopping day of the year. For example, shoppers drove around parking lots for twenty minutes before finding parking just to wait in 30-minute lines to buy their treasured items.

Name ______________________ Date ____________

Definition and Synonym Clues

Sometimes defining an unfamiliar word is easy because its definition, or a familiar synonym for the word, is stated directly in the sentence. **Commas**, **dashes**, and the words **is**, **are**, **was**, **were**, and **or** are often hints that a definition or synonym is coming up. Look for these words and punctuation marks to help you find definitions and synonyms for unfamiliar words in these sentences. Write a definition for the *italicized* word in each sentence.

DEFINITION

1. *Artifacts*, or objects crafted by humans a long time ago, were found at the digging site. ______________________
2. Platte River School, our biggest *rival*, is our scariest competitor. ______________________
3. It is fascinating to study *ancient*, or old, civilizations that no longer exist. ______________________
4. Practicing *deception*—through lying, cheating, or deceiving—will get you kicked out of this school. ______________________
5. *Discrimination*, treating people differently based on race, religion, or gender, will not be practiced here. ______________________
6. Women fought long and hard for *suffrage*, which is the right to vote. ______________________
7. A *replica*—a well-crafted copy—of a Van Gogh is much less expensive than the original. ______________________
8. In earlier times, a salesperson was often called a *merchant*. ______________________
9. I think I can *persuade*, or convince, John to read *The Hatchet*; it's a great book. ______________________
10. An *assembly* is a gathering of people in one place for a common purpose. ______________________
11. We enjoy more civil *liberties*—freedoms guaranteed by law—than do citizens of many other countries. ______________________
12. You will be *obligated*, or committed, to pay back this loan if you take it out. ______________________
13. The U.S. Army has some strict, or *rigid*, rules. ______________________
14. The *fracture*, or break, in his bone will require a cast. ______________________

Name ______________________ Date ____________

Antonym and Contrast Clues

Sometimes readers can define an unknown word by contrasting the unknown word with a known word or idea. Antonym or contrast clues often begin with words such as **unlike**, **as opposed to**, **instead of**, **rather**, and **not**. Use antonym and contrast clues to help you define the italicized words in the passages below. Circle the best definition for each word.

Balloons were first made from animal bladders. Unlike wood or glass, animal bladders have *elasticity*. Rubber does too. Instead of remaining the same dimensions, rubber *expands* when filled with hydrogen or air. In fact, rubber expanded by hydrogen does not become heavy by having something added to it; rather, it becomes so light that it *ascends* into the air.

1. elasticity
   A. transparency **B.** heaviness **C.** ability to stretch **D.** many uses
2. expands
   A. grows **B.** shrinks **C.** explodes **D.** floats
3. ascends
   A. explodes **B.** falls **C.** rises **D.** remains

Some people say technological advancements are not really *advantageous*. Instead of seeing an increase in productivity, we have actually seen a *decline*. Instead of *eliminating* paperwork, technology has created more. Critics even suggest that work done on computers simply *duplicates* work done without them, rather than replacing it.

4. advantageous
   A. contagious **B.** beneficial **C.** having disadvantages **D.** common; wide-spread
5. decline
   A. increase **B.** decrease **C.** improvements **D.** advancements
6. eliminating
   A. simplifying **B.** altering **C.** getting rid of **D.** improving
7. duplicates
   A. doubles **B.** modifies **C.** decreases **D.** contains

Many high schools today require advanced math credits from their graduates, even though the students have already taken basic and *intermediate* level classes. If you don't really like math, you may have to make a *deliberate* effort to take tough math classes anyway, rather than *instinctively* not signing up for them because you don't like math.

8. intermediate
   A. low level **B.** mid-range **C.** highest level **D.** beginning level
9. deliberate
   A. on purpose **B.** accidental **C.** difficult **D.** simple
10. instinctively
   A. selfishly **B.** intentionally **C.** naturally **D.** surprisingly

Name ______________________________ Date ______________

Sentence Structure Clues

Understanding sentence structure helps readers narrow down the definitions of unknown words. Consider this sentence: "The *entozoon* living inside the cow's intestines made it very sick." From its placement in the sentence, you can tell *entozoon* is a noun: a person, place, or thing. Since it is "living inside the cow's intestines" and making the cow sick, it must be a small, living thing that does damage to other living things. Indeed, an *entozoon* is a parasitical, intestinal worm. Use sentence structure clues to help you determine the part of speech and define the italicized words below. Circle the correct definitions.

1. It seems *equitable* that I get to go to the movies Friday night if Jan gets to go tonight.
   A. adjective meaning fair
   B. noun meaning event
   C. verb meaning attend

2. The *antidote* for snakebite is, ironically, snake venom.
   A. verb meaning bite
   B. noun meaning remedy
   C. adjective meaning dangerous

3. The wrecking crew will *demolish* the old building before constructing the new one.
   A. adjective meaning broken down
   B. adverb meaning fiercely
   C. verb meaning destroy

4. My *malicious* sister likes to get me in trouble.
   A. adverb meaning angrily
   B. adjective meaning mean
   C. noun meaning relative

5. We *synchronized* our watches so we could meet up in town at the same time.
   A. adverb meaning together
   B. noun meaning harmony
   C. verb meaning matched

6. The medical *technician* performed my test, but a doctor had to read the results.
   A. noun meaning skilled person
   B. verb meaning perform
   C. adverb meaning quickly

7. Twenty dollars was a great *incentive* to mow the lawn.
   A. noun meaning motivator
   B. verb meaning cut or trim
   C. adverb meaning simply

8. Dad was *incredulous* as he read stories about aliens from outer space.
   A. verb meaning to believe
   B. adjective meaning skeptical
   C. noun meaning spaceship

9. The crowd *boisterously* welcomed home the gold medal Olympian with shouts and cheers.
   A. verb meaning rejoicing
   B. adverb meaning loudly
   C. noun meaning champion

10. My suffering puppy *recoiled* in pain when I tried to bandage his hurt leg.
    A. verb meaning to draw back
    B. noun meaning pain
    C. adjective meaning painful

Name ________________________________ Date ______________

Background Knowledge Clues

Readers use their own experiences and knowledge to help them define unknown words. They may be familiar with an unknown word used in a different context, for example. Or they may know enough about a given subject matter to help them make sense of unfamiliar words within a text. Use your knowledge about Native Americans and any previous contact with the italicized words in the passage to help you define the words. Circle the correct definitions.

The Nez Perce are *native* to the Wallowa Valley of Northeastern Oregon. With the arrival of Lewis and Clark, whites joined them there. At first, the two *cultures* lived *amiably* together. They traded goods and *socialized*. A white missionary even gave one of his native friends an English name. He called his friend "Joseph the Elder."

In 1855, troubles began. The United States government wanted to buy Nez Perce land. Joseph the Elder would not sell it. Twenty-two years later, after Joseph the Elder had died, General Howard of the United States military insisted on taking the land anyway. He gave Joseph the Elder's son, Chief Joseph, and his people thirty days to *relocate*.

Chief Joseph's people did not want to leave their homeland. But they began a long march to a *reservation*. Near the end of their travels, a Nez Perce man killed a white man in *retaliation* for the killing of his father. This ignited a war.

Chief Joseph led his people on a desperate march for freedom in Canada. Eight hundred Nez Perce traveled 1,700 miles across four states. Then the U.S. army caught up with them. A bloody battle followed. Two hundred Nez Perce were killed. It was the dead of winter, and the natives had no food or blankets. Tired of the cold and the killing, Chief Joseph surrendered just 40 miles south of the Canadian border. Those of his people who survived the fighting were marched to a reservation in Oklahoma. Further *negotiations* between Chief Joseph and President Hayes allowed the Nez Perce to move to a reservation in the Northwest, but they never saw their valley again.

1. native
   - **A.** new to a land
   - **B.** immigrants
   - **C.** first to live in a land
   - **D.** foreigners

2. culture
   - **A.** traits of a society
   - **B.** a country
   - **C.** an ethnic group
   - **D.** a family

3. amiably
   - **A.** fighting
   - **B.** in an angry manner
   - **C.** in a friendly manner
   - **D.** silently

4. socialized
   - **A.** visited and talked
   - **B.** yelled and fought
   - **C.** hunted
   - **D.** conducted business

5. relocate
   - **A.** sell property
   - **B.** move to a new place
   - **C.** return to a homeland
   - **D.** plant crops

6. reservation
   - **A.** a hotel
   - **B.** property granted to Native Americans
   - **C.** the countryside
   - **D.** a city

7. retaliation
   - **A.** anger
   - **B.** murder
   - **C.** injury
   - **D.** act done in revenge

8. negotiations
   - **A.** meetings intending to reach agreement
   - **B.** arguments
   - **C.** battles
   - **D.** laws

Name ______________________________ Date ____________

Context Clues Collection

Practice working with your entire collection of context clue skills to determine the meaning of unknown words. Circle the meaning of each italicized word, and write the context clue from the box below that helped you.

CONTEXT CLUES
word part clue
example clue
definition/synonym clue
antonym/contrast clue
sentence structure clue
background knowledge clue

1. Mr. Cader's acts of *philanthropy* included donating both time and money to charities.
   A. charity **B.** wealthy person
   C. popularity **D.** the study of generosity

   Context Clue: ______________________________

2. The envelope's *adhesive* material was like cement; once I sealed it, I couldn't reopen it.
   A. paper **B.** white
   C. sticky **D.** smooth

   Context Clue: ______________________________

3. Sue knew about *famines* from her study of all the deaths from the Potato Famine in Ireland.
   A. wars **B.** scarcity of food **C.** immigration **D.** food source

   Context Clue: ______________________________

4. Unlike the quarter Grandma offered, the five dollars Dad offered was real *motivation* to mow the lawn.
   A. money **B.** chores **C.** reason; cause **D.** punishment

   Context Clue: ______________________________

5. Yes, I'm *irritable* today; everything seems to irritate me.
   A. easy to upset **B.** easy to please **C.** sad **D.** happy

   Context Clue: ______________________________

6. Two hours before take off should leave you *ample* time to check in your luggage at the airport.
   A. one full day **B.** minutes **C.** plenty; enough **D.** insufficient

   Context Clue: ______________________________

7. Kevin made the *economical* decision of buying the reasonably priced shoes.
   A. foolish **B.** wasteful **C.** not wasteful **D.** painful

   Context Clue: ______________________________

8. His position was *justifiable*, backed up by both research and good reasoning.
   A. inexcusable **B.** unusual **C.** unpopular **D.** proven by reason

   Context Clue: ______________________________

Name ______________________________ Date ______________

The Medical World

Use your new understanding of context clues to help you match the definitions below with the medical words italicized here.

1. ______ Heart attack patients see *coronary* doctors.
2. ______ A kidney *transplant* can save a life.
3. ______ Tim was glad his tumor was *benign.*
4. ______ Tonya's *malignant* tumor needed to be removed before the cancer spread.
5. ______ Grandpa's *edema* made it hard for him to fit his feet into his shoes.
6. ______ The bump on Terry's knee is a *tumor.*
7. ______ Your heart *valves* control your blood flow.
8. ______ Kim's *irregular* heartbeat is unusual but not dangerous.
9. ______ Symptoms of *angina* may cause your doctor to suspect a heart attack.
10. ______ *Plaque* will build up on your teeth if you don't brush them.
11. ______ Heart doctors need to study *cardiology.*
12. ______ Therapists study *psychology.*

DEFINITIONS

A. swelling
B. having to do with the heart
C. not cancerous
D. not normal
E. bacteria deposits
F. cancerous
G. abnormal growth
H. devices for opening and closing
I. chest pains
J. study of the heart
K. to transfer an organ
L. study of the mind

Name ______________________________ Date ______________

Grasping Government

Use context clues to help you define the italicized words in the sentences below. Circle the correct definitions.

1. The *government* of the United States consists of three branches that work together to run the country.
   A. White House
   B. democracy
   C. a system that controls public affairs
   D. president

2. The *governing* body of our school is the Central County School Board.
   A. controlling public affairs
   B. taking notes
   C. voting
   D. teaching

3. Wills, leases, contracts, and birth certificates are legal *documents*.
   A. agreements
   B. official papers
   C. books
   D. laws

4. Our building *administrator* makes decisions, fills out work orders, and sees to it that things run smoothly.
   A. secretary
   B. manager
   C. janitor
   D. teacher

5. The store's *policy* does not allow returned items without a receipt.
   A. rules
   B. director
   C. computer
   D. sales clerk

6. My teacher has the *authority* to give out detentions to misbehaving students.
   A. personality
   B. anger
   C. official power
   D. desire

7. My dad follows *politics* more closely in election years.
   A. the practice of governing
   B. senators
   C. history
   D. the news

8. Oregon's state minimum wage is higher than the *federal* minimum wage.
   A. of one state
   B. of one town
   C. of one county
   D. of the entire country

9. Senator Dample used several clever *tactics* to get that bill passed.
   A. advertisements
   B. luncheons
   C. money managers
   D. methods

10. France's *constitution* contains many of the same rights as does ours.
    A. senate
    B. population
    C. system of laws and rights
    D. Bill of Rights

Name ______________________________ Date ______________

Interpreting Action

Use context clues to help you match the verbs in italics with their definitions.

1. ______ The rock concert was *teeming* with screaming teens in jeans and T-shirts.
2. ______ The sheet hanging on the clothesline *billowed* in the breeze.
3. ______ The hot sun *scorched* my green grass, turning it dry and brown.
4. ______ After swimming all day on an empty stomach, I *devoured* my dinner.
5. ______ Reading the great review *influenced* Gabby's decision to see the movie.
6. ______ The wet metal of the slide on the playground *glistened* in the sun.
7. ______ I didn't mean to *offend* you by forgetting to send a thank-you card.
8. ______ I hope you can *pardon* my mistake and give me another chance.
9. ______ I would like to *inquire* about the car you have for sale, because I'm looking for one.
10. ______ Reading the entire story *quenched* my curiosity about the article's strange title.
11. ______ Dad will likely *consent* to your going to the show tonight; just ask him.
12. ______ Since the doctor said I may be low on iron, I'll *adjust* my diet to include raisins.

DEFINITIONS

A. shimmered and shined
B. crowded
C. had an affect on
D. change; modify
E. ask
F. forgive
G. give permission
H. hurt someone's feelings
I. swayed and flopped
J. relieved a thirst or desire
K. ate greedily
L. burned

Name ______________________________ Date ______________

Plenty of Personality

Use context clues to help you fill in each blank with the correct word related to personality and emotions.

WORD BOX

lavish
perfectionist
magnetic
tiresome
insistent
distress
husky
treacherous
reliable
aggression
explosive
reserved

1. Larry, my boring uncle, tells the same ______________ stories over and over again.

2. Jenny's ______________ temper gets her in trouble with the referees of her soccer games.

3. Sadie is such a ______________ she becomes upset at missing a single question on a test or homework assignment.

4. My bossy sister is ______________ on getting her way.

5. Karen's tears revealed her ______________ when her dog ran away.

6. My shy, modest, and ______________ grandmother would never join in karaoke night with us.

7. Kevin, the daredevil, hiked some ______________ backcountry paths through the Rocky Mountains last summer.

8. You can depend on Cindy; she is a ______________ person.

9. Unlike his well-mannered behavior in the classroom, Ari displays ______________ on the basketball court.

10. Tim's deep, ______________ voice reminds me of his father's voice.

11. One look at all of Roman's expensive home furnishings will convince you of his ______________ taste in decorating.

12. Nelle's ______________ personality really does attract people like a magnet.

Name ______________________________ Date ______________

Visual Vocabulary

Below are sentences containing italicized words from the visual arts. Use context clues to help you decide whether the statements following the sentences are true or false. Circle **T** for true or **F** for false.

T F 1. Looking closely at the *mosaic*, I could see all the tiny pieces of glass that comprised its complete image.

➧ A *mosaic* is a unified design that is made of individual tiny pieces.

T F 2. Stacy's grandmother painstakingly *embroidered* flowers onto my pillowcase one stitch at a time.

➧ When Stacy's grandmother *embroidered* the pillowcase, she painted it.

T F 3. I thought I'd never finish stenciling the owl pattern onto the silk for my *screenprint.*

➧ A *screenprint* is made by forcing ink through patterns in some type of material.

T F 4. The sunlight creates just enough shadow to *silhouette* your image against the side of that building.

➧ A *silhouette* is a dark shape against a light background.

T F 5. Glass and metal are both good surfaces for *etching.*

➧ *Scratching* is a synonym for *etching.*

T F 6. After you shape your clay, place it in the *kiln* to create your own cup.

➧ A *kiln* is used to paint clay.

T F 7. If you shape your cup on the *potter's wheel*, it will turn out well rounded.

➧ A *potter's wheel* would most likely be used by a painter.

T F 8. You could also decide to *sculpt* your cup by hand if you want it to look natural.

➧ To *sculpt* means to shape with your hands.

T F 9. Painters' *instruments* include brushes, paints, and easels.

➧ *Instruments* always make music.

T F 10. Add *texture* to your collage with beads, ribbons, and tiles.

➧ The *texture* of an item can be determined by touch.

T F 11. A lighter *hue* in the sky might make your painting brighter.

➧ A synonym for *hue* is *shade.*

T F 12. Sculpting is a more *tactile* art than is painting.

➧ The word *tactile* must be related to touch.

Name ______________________________ Date ______________

Board Game Bonanza

Use context clues to help you define the italicized words in the passages below. Circle the correct definitions.

Checkers is a two-person game played with black and red pieces on a board of *alternating* black and red squares. Its *strategy* is based on attacks and *counterattacks*. It is an ancient game that is thought to be a *precursor* of chess.

1. alternating
   A. following in turns **B.** both **C.** several **D.** few
2. strategy
   A. rules **B.** game **C.** competition **D.** plan of action
3. counterattacks
   A. the original move **B.** planning **C.** cooperation **D.** moves following a move
4. precursor
   A. copy **B.** forerunner **C.** similar **D.** unrelated

Chess is thought to have *originated* in India. It features sixteen pieces that can be *maneuvered* in various ways on a checkered board. *Publicized* world championship matches *promote* the game and give it *prestige*.

5. originated
   A. started **B.** stayed **C.** become popular **D.** left; abandoned
6. maneuvered
   A. balanced **B.** moved **C.** jumped **D.** placed
7. publicized
   A. long **B.** popular **C.** advertised **D.** skilled
8. promote
   A. make popular **B.** change **C.** downgrade **D.** explain
9. prestige
   A. popularity **B.** respect **C.** competitiveness **D.** followers

Bobby Fischer became world famous when he gained *international* grand master *status* as a chess player at the age of 15. In 1972, he became the world chess champion when he defeated Boris Spassky.

10. international (between many…)
    A. states **B.** countries **C.** cities **D.** counties
11. status
    A. trophy **B.** friends **C.** standing; position **D.** competitors

Name ______________________________ Date ______________

What's with the Weather?

Use context clues, including your understanding of sentence structure, to help you fill in the blanks in the passages below with weather words from the boxes.

A The ____________(1) temperature kept most people indoors today. But Bobby is out ____________(2) up Riverette Mountain. No ____________(3) weather is going to keep him from his favorite winter ____________(4). I'm not sure an ____________(5) tumbling down the mountainside could even slow him down.

WORD BOX
recreation
inclement
snowshoeing
avalanche
frigid

B Visiting Thailand last year was wonderful, but the ____________(6) temperatures and heavy rainstorms reminded us it was ____________(7) season. The ____________(8) predicted no ____________(9) during our visit, so the heavy storms did not worry us much.

WORD BOX
typhoon
monsoon
meteorologist
sweltering

C The ____________(10) bright sun and ____________(11) heat at the beach yesterday were almost enough to make me go home early. But frequent dips into the ocean ____________(12) me, and a rich ____________(13) of sunscreen kept me from burning.

WORD BOX
sultry
refreshed
blazingly
slathering

Name ______________________________ Date ______________

Community Service

Use context clues to help you define the italicized words in the reading. Circle the correct definitions.

My history teacher, Mr. Malone, doesn't like schoolbooks much. He delights in projects. So when we began a study of community, our whole class walked to City Hall to meet the mayor. Then we were *spectators* at a trial at the courthouse next door. On the way back to school, we stopped at the city park and picked up garbage. Then, Mr. Malone assigned an *individual* project. We had three weeks to do something *significant* for our community and report back to class. "A *community* is a collection of all its members," Mr. Malone explained, "and we must all work for the community to work."

All the project lovers were excited. I wasn't! In fact, I *procrastinated* until the day before the three weeks were up. Then I begged Mom for an idea. She suggested I ask Mrs. Crumsky if she needed help with any chores. Mrs. Crumsky lives next door. She's about a hundred years old, but she didn't want any help with her chores. She just wanted to talk. And she did—for three hours!

The next day in class, I was sweating bullets when Mr. Malone called on me to share my project. "I'm sorry, Sir," I squeaked. "I have nothing to report. I offered to help my hundred-year-old neighbor with her chores, but all she wanted to do was talk."

Mr. Malone *smirked* and *extracted* a letter from his jacket pocket. He began to read, "Dear Mr. Malone, The newspaper says your class is completing a community project. I suppose that is why Charlie asked me if I needed help with any of my chores. I told Charlie I do all of my own chores, and I do. That poor boy must think he didn't *accomplish* much. But I want you to know he did, Mr. Malone. Charlie let me talk his ear off for three hours. Community is about people needing other people, and I needed that talk more than I even knew. I'm only one old lady in this town, but Charlie brought sunshine to my day, and I think that's worth an A+. Sincerely, Mrs. Crumsky."

1. spectators
   A. intruders
   B. observers
   C. students
   D. jury members

2. individual
   A. group
   B. relating to children
   C. relating to a single person
   D. service

3. significant
   A. important
   B. lengthy
   C. complicated
   D. involving service

4. community
   A. service projects
   B. a governor
   C. a neighborhood of people
   D. business partners

5. procrastinated
   A. complained
   B. completed early
   C. did not complete
   D. put off until the last minute

6. smirked
   A. frowned
   B. laughed
   C. smiled knowingly
   D. nodded a head

7. extracted
   A. removed
   B. created
   C. produced
   D. read

8. accomplish
   A. to listen
   B. to do physical work
   C. to achieve
   D. to waste time

Name ______________________________ Date ______________

Space-tacular Words

Use context clues to help you define the italicized space-related words. Circle the correct definitions.

1. The unmanned *lunar* vehicle sampled moon rocks.
   - **A.** truck-like
   - **B.** equipment designed for experimenting
   - **C.** relating to the moon
   - **D.** sun

2. Eight planets *orbit* the sun.
   - **A.** stay close to
   - **B.** revolve around
   - **C.** move in a straight line
   - **D.** remain still

3. The Big Dipper is my favorite *constellation.*
   - **A.** cluster of stars
   - **B.** planet
   - **C.** comet
   - **D.** star

4. The Milky Way is not the only *galaxy*; countless other star systems exist.
   - **A.** star
   - **B.** planet
   - **C.** collection of stars, gas, and dust
   - **D.** comet

5. *Satellites* in space send signals for television, radio, and cell phones.
   - **A.** airplanes
   - **B.** space stations
   - **C.** spaceships
   - **D.** objects that orbit the earth

6. Signals bounce off the satellites and are *transmitted* in a desired direction.
   - **A.** sent or passed on
   - **B.** electricity
   - **C.** received
   - **D.** heard

7. The glittery *luster* from your diamond nearly matches that of the stars.
   - **A.** body
   - **B.** shine
   - **C.** curl
   - **D.** perfume

8. Tonight's full moon provides almost enough *illumination* to allow me to see where I am driving without my car lights on.
   - **A.** excitement
   - **B.** light
   - **C.** heat
   - **D.** distraction

9. The Milky Way is only one galaxy in the *universe.*
   - **A.** planet
   - **B.** star
   - **C.** everything that exists
   - **D.** moon

10. *Interstellar* gases are as much a part of the universe as the stars themselves.
    - **A.** moons
    - **B.** bright lights
    - **C.** odorous
    - **D.** among the stars

Name ______________________ Date ______________

The Generals Who Became Emperors

Use context clues to help you define the italicized words in the passages below. Circle the correct definitions.

Julius Caesar and Napoleon Bonaparte were *renowned* for their military skills. As young and *ambitious* generals, they both conquered much of Europe. Their armies moved with lightning *rapidity* and overwhelming force.

1. renowned
   A. famous **B.** infamous **C.** unknown **D.** disliked
2. ambitious
   A. lazy **B.** eager **C.** popular **D.** powerful
3. rapidity
   A. speed **B.** intelligence **C.** anger **D.** determination

Neither general was *content* with simple *conquest*. Caesar and Napoleon *expended* great effort on improving government and laws in the lands over which they *reigned*. Europe today still shows their remarkable influence.

4. content
   A. unhappy **B.** satisfied **C.** to charge, as into battle **D.** in disagreement
5. conquest
   A. ruling **B.** power **C.** fame **D.** victory
6. expended
   A. expected **B.** demanded **C.** given out **D.** expanded
7. reigned
   A. conquered **B.** lived **C.** ruled **D.** fought

Both generals declared themselves *emperor*. Although they enjoyed great *mainstream* popularity, some people considered them *politically* dangerous. Caesar was *assassinated*. Napoleon was defeated and *exiled* to an island, where he died.

8. emperor
   A. sergeant **B.** military leader **C.** ruler of a city **D.** ruler of an empire
9. mainstream
   A. moderate **B.** great **C.** widespread **D.** underground
10. politically (having to do with…)
    A. war **B.** the economy **C.** society **D.** governing
11. assassinated
    A. killed by a relative **B.** killed for political reasons **C.** punished **D.** killed by accident
12. exiled
    A. removed from home country **B.** vacation **C.** travel **D.** govern

Name ______________________________ Date ______________

How Do They Do That?

Use context clues to help you fill in the blanks in the sentences below with adverbs from the Word Box.

1. I never wear a dress. I ____________________ don't even own one!
2. I ate my dinner ____________________ after running the marathon.
3. She reads for an hour before bed ____________________; she never misses a day!
4. Working out ____________________ for 45 minutes a day is a great way to keep in shape.
5. I looked ____________________ for my wallet when I lost it. It had $100 in it!
6. I am ____________________ amused by slapstick humor, but I'd rather listen to funny dialogue.
7. Although I like to browse, I ____________________ buy anything at the crafters' market.
8. Jordan ____________________ attends every high school sporting event in town.
9. I ____________________ vowed to uphold the principles of the honorable organization.
10. I'm flying back to Indiana because my grandfather is ____________________ ill.
11. Many members of our football team this year are not only athletically excellent but ____________________ successful as well.
12. I would not tell this to anyone except you, but ____________________, I don't really like Grandma's bread pudding.

WORD BOX

heartily
mildly
literally
fanatically
scholastically
religiously
vigorously
frankly
seldom
gravely
solemnly
frantically

Name ______________________ Date ____________

Water Words

Use context clues to help you match the italicized words with their definitions.

1. ______ I put my contact lenses in an *aqueous* solution at night.
2. ______ The oceans are filled with both plant and animal *marine* species.
3. ______ Water polo is the *aquatic* version of land polo.
4. ______ Derry's boat is anchored at the *marina*.
5. ______ Drive slowly so you don't *hydroplane* in this rain.
6. ______ Using *hydroponics*, Jenny needs no soil for her tomato crop.
7. ______ The Hoover Dam provides an important source of water for *hydroelectricity*.
8. ______ Replace the fluid often in your *hydraulic* lift for best performance.
9. ______ Zeke enjoyed viewing the ships at the *maritime* museum.
10. ______ In the *Odyssey*, Odysseus and the other *mariners* get lost at sea.
11. ______ Drink water to stay *hydrated* while playing sports.
12. ______ Linda, who suffers from *hydrophobia*, will not go near the ocean!

DEFINITIONS

A. electricity produced from fast-moving water
B. relating to the sea
C. to glide along on the surface of water
D. relating to human activity at sea
E. taking place in the water
F. growing crops in water and chemicals
G. a port or parking lot for pleasure boats
H. operating using the pressure of a fluid
I. fear of water
J. sailors
K. water-based
L. to supply water to maintain fluid balance

Name ____________________ Date ____________

Preferring Prefixes

Use context clues, including your understanding of word parts, to help you define the italicized words. Circle the correct definitions.

1. Allen's *overactive* thyroid gland makes him lose weight no matter how much he eats.
   - **A.** sluggish
   - **B.** inactive
   - **C.** too active
   - **D.** diseased

2. Zeke's *disarming* smile melted away his mother's angry glare after he spilled the milk.
   - **A.** funny
   - **B.** removing weapons or defenses
   - **C.** sad
   - **D.** trick

3. Eric was *discharged* from the hospital after his successful operation.
   - **A.** allowed to leave
   - **B.** scheduled to return
   - **C.** visited
   - **D.** required to stay

4. Eric was doing well after his surgery, but he has now suffered a *relapse.*
   - **A.** fever
   - **B.** new illness
   - **C.** new injury
   - **D.** recurrence of an illness

5. Tracy played a practical joke on me in *reprisal* for my April Fool's Day performance.
   - **A.** retaliation
   - **B.** honor of
   - **C.** humor
   - **D.** rememberance of

6. I'll need to *replenish* our supplies if we camp an extra week.
   - **A.** consume
   - **B.** resupply
   - **C.** hide
   - **D.** reserve

7. That is a flawless *reproduction* of Van Gogh's masterpiece.
   - **A.** painting
   - **B.** copy
   - **C.** original
   - **D.** authentic

8. The Diver County Court decision upholding curfew laws for teens set a *precedent* that other counties in the state soon followed.
   - **A.** judge
   - **B.** law
   - **C.** early example
   - **D.** jury

9. I was *preoccupied* with my homework and didn't hear what you said.
   - **A.** annoyed
   - **B.** ignored
   - **C.** completing
   - **D.** distracted by preexisting thoughts

10. I vary my exercise routine, but I am *inflexible* in my commitment to work out 30 minutes a day.
    - **A.** unconcerned
    - **B.** concerned
    - **C.** not flexible
    - **D.** very flexible

Name ______________________________ Date ______________

Mr. Fribble Learns Too Late

A Modern Retell of "The Cop and the Anthem" by O. Henry

Use context clues to help you define the italicized words in the reading. Circle the correct definitions.

Mr. Fribble was a kind old man, forced into retirement by his *advancing* years that slowed him down on the job. Tomorrow, he would be out of work and out of money. On his *meager* retirement pay, he would be out of a home, too. But Mr. Fribble had a plan. He would land himself in jail so he would have a warm place to live. Mr. Fribble tried everything to get arrested. He dined in an *extravagant* restaurant without paying, but he was just made to wash dishes. He *snatched* a purse from a gray-haired lady, but she just hit him on the head with her cane until he gave it back.

Dispirited, he sat down to think at the corner of Fifth and Main. A children's choir sang inside a church on that corner, and Mr. Fribble stared in the window. As he listened to the soft sounds of music, a stray dog cozied up to Mr. Fribble and listened too. At that moment, Mr. Fribble decided not to go to jail after all. The music *inspired* him to make plans for a new start! He would find a new job where his *ponderous* pace wouldn't matter. But just as Mr. Fribble turned to grab the help wanted ads from a paper stand behind him, he found himself facing a pair of handcuffs. "Sir, dogs in this city must remain on a leash. You'll have to come with me," said a police officer. "But, but..." Mr. Fribble *stammered*, as he was led directly to jail.

1. advancing
   - **A.** young
   - **B.** moving forward
   - **C.** numerous
   - **D.** slowing down

2. meager
   - **A.** not enough
   - **B.** monthly
   - **C.** full
   - **D.** abundant

3. extravagant
   - **A.** inexpensive
   - **B.** busy
   - **C.** fancy
   - **D.** delicious

4. snatched
   - **A.** looked at
   - **B.** requested
   - **C.** grabbed
   - **D.** bargained for

5. dispirited
   - **A.** discouraged; depressed
   - **B.** exhausted
   - **C.** angry
   - **D.** encouraged; hopeful

6. inspired
   - **A.** changed one's mind
   - **B.** loud
   - **C.** motivated and encouraged
   - **D.** performed

7. ponderous
   - **A.** quick
   - **B.** inconsistent
   - **C.** moderate
   - **D.** slow and thoughtful

8. stammered
   - **A.** yelled
   - **B.** whispered
   - **C.** cried
   - **D.** stuttered

Name ______________________________ Date ______________

A Job Well Done—or Otherwise

Use context clues to help you decide which synonym from below best replaces the italicized word related to work in the sentences that follow.

1. ______ Ruth did a *meticulous* job of cleaning her room; not a speck of dirt remains.
2. ______ Ruth's sister did a more *dubious* job of cleaning her room. In fact, I couldn't actually tell she had cleaned, even after she had finished.
3. ______ Thomas Edison *doggedly* tried over a thousand materials before he found one that would burn bright and long in lightbulbs.
4. ______ I make my appointments with David because he is a *competent* hairdresser.
5. ______ Your outdated computer is *inadequate* to complete the graphic work this project will require.
6. ______ I selected Dr. Jones because his *professionalism* impressed me when he worked on my sister's teeth.
7. ______ It is *urgent* that I get this project turned in on time.
8. ______ Veterinarians are *reluctant* to work on people, but their training does prepare them to do so in case of emergencies.
9. ______ You can *accelerate* your program and graduate early.
10. ______ *Employing* only a paring knife, the chef transformed common vegetables into beautiful flowers.
11. ______ Katie is a *diligent* worker who comes in early and frequently stays late.
12. ______ Pitting cherries is a *tedious* job that requires the same motion over and over.

SYNONYMS

A. able
B. using
C. hardworking
D. boring
E. questionable
F. stubbornly
G. careful
H. incapable
I. rush
J. hesitant
K. expertise
L. crucial

Name ____________________ Date ____________

A Display of Attitude

Use context clues to help you fill in the blanks with the appropriate words about attitude from the box below.

WORD BOX

opinionated
confident
collective
jovial
arrogant
manipulated
domineering
congenial
prejudice
submissive
insecure
assertiveness

1. My ____________________, or controlling, boss allowed no one else to make a decision.

2. Sasha is a ____________________ person who believes she will accomplish anything she sets out to do.

3. The community garden is a ____________________ project, requiring the help of all the neighbors in the subdivision.

4. Tanya's appreciation of her little sister's friend made her rethink her ____________________ against anyone under the age of 12.

5. Rachel's ____________________, or pleasant, personality made her enjoyable to be around.

6. My ____________________ friend always thinks she has done poorly on her tests, even though she always does fine.

7. My ____________________ dog will obey any command you give him.

8. Charles is so ____________________ that there is not a subject he doesn't have an opinion about.

9. Santa Claus is known for his ____________________ spirit and his great white beard.

10. Maurico became ____________________ after winning the spelling bee, making it hard for his friends to be happy with him.

11. Sandra enrolled in ____________________ training to help her learn to state her needs clearly.

12. Tyler ____________________ his friends into bringing him candy bars by convincing them his doctor prescribed chocolate for his cold.

Name ______________________________ Date ______________

Talk About Time

Use context clues to help you define the italicized words pertaining to time. Circle the correct definitions.

1. Mr. Randel has a *chronic* backache that does not respond to treatment but continues to hurt.
   A. severe
   B. persistent
   C. temporary
   D. medical

2. Mrs. Henson told us we could read the novels in her classroom at our *leisure*.
   A. free time
   B. time of an appointment
   C. requirement
   D. never

3. A marathon is so long that it takes a great deal of *stamina* to complete it.
   A. breath
   B. hydration
   C. carbohydrates
   D. perseverance

4. *Contemporary* stories are easier to read since they use current vocabulary and styles.
   A. easy
   B. modern
   C. classic
   D. old

5. I can't remember the last time we went to the movies; it has been *aeons*.
   A. remains the same
   B. a long time
   C. never occurs
   D. ten years

6. *Perennial* flowers are planted once and blossom year after year.
   A. occurring once
   B. occurring twice
   C. recurring every year
   D. occurring unpredictably

7. If you do not hurry up with step two, you will *lag* behind on this project.
   A. constantly change
   B. constantly move
   C. become delayed
   D. get ahead

8. Can you name the presidents in *chronological* order beginning with Washington?
   A. alphabetical order
   B. oldest to youngest
   C. time order
   D. least to most popular

9. You can shop at both because the craft market and fruit market events will be going on *simultaneously*.
   A. at different times
   B. one after the other
   C. at the same location
   D. at the same time

10. My stick-on tattoo is *temporary*, but my dad's is permanent.
    A. lasting a long time
    B. lasting forever
    C. not permanent
    D. fake

Name ______________________________ Date ______________

The Ant

Use context clues to help you define the italicized words in the passages below. Circle the correct definitions.

Ants live in *colonies* that are something like the cities that people *inhabit*. Like people, ants *communicate* with various *techniques*. They see, hear, feel, taste, and smell messages from other ants.

1. colonies
   A. ant hills **B.** homes **C.** communities **D.** structures
2. inhabit
   A. build in **B.** live in **C.** visit **D.** work in
3. communicate
   A. share information **B.** travel **C.** listen **D.** follow instructions
4. techniques
   A. talents **B.** chores **C.** tools **D.** ways of doing things

Ants in a colony are not all equal. They belong to different *castes*, with different duties. Every ant has a *vocation*. Some are workers, some are soldiers, some watch over the new *larvae*, and some are queens.

5. castes
   A. levels **B.** families **C.** communities **D.** leaders
6. vocation
   A. tool **B.** job **C.** favorite vacation spot **D.** personality
7. larvae
   A. newly hatched insect **B.** eggs **C.** food supply **D.** adult form of insect

Usually ants and people *co-exist* peacefully. Because ants eat other pests, their colonies can be *beneficial* to people. However, red fire ants have painful bites, and carpenter ants can *demolish* wooden buildings.

8. co-exist
   A. fight **B.** live in peace **C.** socialize **D.** ignore each other
9. beneficial
   A. annoying **B.** favorable **C.** destructive **D.** in the way of
10. demolish
   A. destroy **B.** preserve **C.** live in **D.** eat

Name ______________________________ Date ______________

Eatin' Up Vocabulary

Use context clues to help you match the italicized food-related words with their definitions.

1. ______ No one can match my dad's *savory* grilled T-bones with the special sauce.

2. ______ Although onions are *malodorous* to me, Mom loves their smell.

3. ______ Mom, who loves everything from Brie to blue, is a cheese *aficionado.*

4. ______ She even likes the *pungent* smell of a cheese factory!

5. ______ *Sauté* the vegetables and the strips of beef in the same skillet.

6. ______ Sugar and flour are *staples* in the cupboard of a baker.

7. ______ Add brown rice or potatoes to your meal so it will include a healthy *starch.*

8. ______ At the breakfast buffet, both egg dishes were served in hot trays at the buffet table, and cereals were offered in *dispensers* beside the table.

9. ______ See if they offer popcorn at the *concession* stand.

10. ______ I don't use *utensils* when eating pizza; I just pick it up.

11. ______ The preparatory chef had an extensive collection of *cutlery.*

12. ______ You *grate* the cheese, I'll slice the pepperoni, and we'll get this pizza in the oven.

DEFINITIONS

A. eating tools
B. special flavor
C. carbohydrate
D. shred
E. knives
F. sharp odor or taste
G. enthusiast; fan
H. basic, common product
I. fry quickly
J. bad smelling
K. place of business; especially for the sale of food
L. serving containers

Name ______________________________ Date ______________

Just Playin' Around

Use context clues to help you place words from the box into these sentences pertaining to sports.

WORD BOX

division
swerved
regulations
equipped
darted
distinguished
participate
dismay
collisions
astounding
scheme
standard

1. Jim ____________________ to avoid the misplaced cone in the motorcross track.

2. Race tracks are designed so that fast-moving cars can avoid ____________________.

3. One of our school's earliest baseball coaches, now 82 years old, was a ____________________ guest of honor at our sports awards night.

4. Much to his ____________________, Clinton narrowly missed the goal.

5. Randy's cheating ____________________ was discovered before the game even began, so he was benched for the day.

6. We use the school's unmarked basketballs for games because they meet league ____________________.

7. It is a ____________________ requirement that all athletes pass a physical before practicing with the team.

8. This van is ____________________ with enough seats to transport the entire team to away games.

9. Susie rebounded the ball, ____________________ down the court, and made a basket before anyone could catch up to her.

10. In order to ____________________ in organized sports, you need to know how to cooperate and follow rules.

11. We play every 2-A ____________________ team in our region during each season.

12. Charlie's ability to guard Colton is ____________________ considering Charlie's height of 4'9" and Colton's height of 5'7".

Name ______________________________ Date ______________

Adventure in New Lands

Use context clues to help you define the italicized words in the passages about exploration. Circle the correct definitions.

The spirit of *exploration* is at the *foundation* of the West. Lewis and Clark first led early explorers on a two-year *expedition* of the Louisiana Purchase Territory in the early 1800s. *Countless* homesteaders followed their lead.

1. exploration (study of...)
   A. a new place **B.** space **C.** the West **D.** plants
2. foundation
   A. location **B.** basis **C.** geology **D.** biology
3. expedition
   A. camping trip **B.** unplanned vacation **C.** hike **D.** journey with a goal
4. countless
   A. few **B.** a dozen **C.** too many to count **D.** none

One and a half *centuries* later, space explorers continued the *tradition*. The United States and Russia began a space race with the *launching* of satellites in the 1950s. In 1969, a *manned* spacecraft landed on the moon.

5. century
   A. five years **B.** ten years **C.** fifty years **D.** one hundred years
6. tradition
   A. custom **B.** space exploration **C.** popular item **D.** travel
7. launching
   A. propelling **B.** building **C.** buying **D.** selling
8. manned
   A. operated by humans **B.** expensive **C.** run by remote control **D.** full

The Explorer's Club, *founded* in 1904 in New York City, *reveres* exploration. Club members explore the globe—and space—and provide the public with lectures, *publications*, and informational programs that promote exploration. They even host an *annual* dinner where strange bugs and plants one might eat in the wild are served.

9. founded
   A. discovered **B.** started **C.** historically **D.** classical
10. reveres
    A. discourages **B.** creates **C.** honors **D.** dislikes
11. publications
    A. published materials **B.** DVDs **C.** videos **D.** radio broadcasts
12. annual
    A. daily **B.** weekly **C.** monthly **D.** yearly

Name ______________________________ Date ______________

Now That's Entertainment

Use context clues to help you define the italicized words from the world of entertainment. Circle the correct definitions.

1. The *screenplay* of that movie was written by a famous author.
   - **A.** movie based on a novel
   - **B.** novel based on a movie
   - **C.** script
   - **D.** short story

2. Mr. Henson's class put on a marvelous *production* of a popular play.
   - **A.** song
   - **B.** storytelling contest
   - **C.** creation
   - **D.** artwork

3. There was not a famous name in the entire *cast* of that movie.
   - **A.** directing staff
   - **B.** production staff
   - **C.** actors
   - **D.** screenplay authors

4. The *choreography* of that fight scene featured several ballet moves.
   - **A.** planned movements
   - **B.** unrehearsed dancing
   - **C.** ballet dancing
   - **D.** gymnastics

5. The audience was so impressed with Sara's performance that it gave her a standing *ovation.*
   - **A.** excited applause
   - **B.** thank you notes
   - **C.** roses thrown at a stage
   - **D.** smiles

6. Sit quietly for the first half of the show, and I'll get you candy at *intermission.*
   - **A.** a candy store
   - **B.** theater lobby
   - **C.** break in a performance
   - **D.** end of a show

7. Paul played beautifully at his piano *recital.*
   - **A.** movie soundtrack
   - **B.** live play
   - **C.** music video
   - **D.** live performance

8. Since J. K. Rowling finished the *manuscript,* her new book will be out soon.
   - **A.** text of a story
   - **B.** idea
   - **C.** latest movie
   - **D.** character development

9. The *orchestra* played classical melodies on classic instruments.
   - **A.** rock band
   - **B.** group of musicians
   - **C.** soloist
   - **D.** three musicians playing together

10. That solo is meant to be sung without *accompaniment* from any instrument.
    - **A.** harmony
    - **B.** printed music
    - **C.** instrumental support
    - **D.** strict rhythms

Name ______________________ Date ____________

Challenges of Nature

Use context clues to help you define the italicized words in the passages below. Circle the correct definitions.

Mt. Everest is the world's tallest mountain and the most famous. Its *geological* age is young, so it hasn't been *eroded* down. It soars five and a half miles high. Temperatures can *plummet* to a hundred degrees below zero, while winds howl at 120 miles per hour.

1. geological (study of...)
   A. dinosaur era **B.** earth's history **C.** weather **D.** mountains
2. eroded
   A. explored **B.** totally demolished **C.** worn away by weather **D.** knocked over
3. plummet
   A. move to **B.** become **C.** rest at **D.** fall sharply

The Amazon is the world's mightiest river. Wildlife *proliferates* in the Amazon. In its waters can be found twenty-foot-long black caiman crocodiles, thirty-foot *anacondas*, and that *carnivorous* fish, the *notorious* piranha.

4. proliferates
   A. increases rapidly **B.** decreases rapidly **C.** eats **D.** survives
5. anacondas
   A. water animals **B.** spiders **C.** mammals **D.** snakes
6. carnivorous
   A. large **B.** small **C.** plant-eating **D.** meat-eating
7. notorious
   A. honored **B.** infamous **C.** enormous **D.** hungry

Siberia *constitutes* more than half of Russia. Much of it is *tundra*, where the ground is frozen solid all year. Siberia is *sparsely* populated. Temperatures drop to an *insufferable* 90° F below zero. In winter, rivers are so thickly iced that trucks use them for roads.

8. constitutes
   A. rules **B.** populates **C.** differs from **D.** makes up
9. tundra
   A. warm, tropic region **B.** frozen, treeless plane **C.** thick forest **D.** unexplored area
10. sparsely
    A. heavily **B.** many animals **C.** many humans **D.** lightly
11. insufferable
    A. unable to tolerate **B.** unable to change **C.** freezing **D.** amazing

Name ______________________________ Date ______________

Dozens of Disciplines

Use context clues to decide whether the statements about the italicized words are true or false. Circle **T** for true or **F** for false.

T F 1. There are dozens of *disciplines* to choose from at college.
➧ Discipline used in this sense must refer to branches of knowledge.

T F 2. After my cousin completed *culinary* school, he became a chef.
➧ A person interested in becoming a nurse should attend culinary school, too.

T F 3. Ted knows a lot about *psychology*, but he's still crazy.
➧ Psychology is the study of the mind.

T F 4. One *philosophy* of life suggests everything happens for a reason.
➧ The word philosophy must refer to the one best way to understand life.

T F 5. Todd was smart enough to attend college, but *economic* troubles kept him out.
➧ This sentence suggests it costs a lot of money to go to college.

T F 6. The word *media* doesn't only refer to newspapers and television; it includes many forms of communication.
➧ Most people would consider statues of famous people part of the media.

T F 7. If you want to own your own company, take some *business* classes first.
➧ A business class would probably cover information on selling and budgeting.

T F 8. Computer *technology* changes so quickly that a year-old machine is ancient.
➧ In this sentence, technology refers to electronic achievements.

T F 9. The *radiologist* will look at your x-ray to see if your arm is broken.
➧ Radiology is the study of blood diseases.

T F 10. Social is the root word for socialize and *sociology*.
➧ Since -ology means "the study of," sociology means the study of how people interact.

T F 11. If you've studied child *development*, you'd know Tyler isn't expected to walk yet.
➧ Development refers to changes over time.

T F 12. Before you join the police department, you will need to take a class in *criminology*.
➧ A class in criminology would include a study of criminal behavior.

Name ______________________________ Date ______________

Troubleshooting Technology

Use context clues to help you complete the sentences with computer jargon from the word box.

WORD BOX

troubleshoot
hacker
font
blog
banners
graphics
document
cyberspace
search engine
home page
formatting
insert

1. Mrs. Heffelfinger always requires us to use 12-point __________________ for our papers.

2. There is something wrong with my monitor. I can hear the sound and read the text, but I can't see any __________________.

3. A __________________ changed the code in my word processing program so my caps lock now turns capital letters off instead of on.

4. If you __________________ the problem, including checking your outlets and switches, maybe you will find out what the problem is with your printer.

5. You can travel around in __________________ without leaving the chair in front of your computer.

6. Blinking advertisement __________________ distract me when I'm reading news on the Web.

7. I've set our school's Web site as my __________________, so I see it everytime I get online.

8. Once you learn to use a __________________ effectively, you can research anything with the right keywords.

9. I'll be done with my paper as soon as I complete the __________________ changes my teacher requested, including the addition of bullets by my seven main points.

10. Be sure to __________________ a page break after each page so your document prints out correctly.

11. I would like both a hard copy and a disk copy of your __________________.

12. I posted a comment on Tammy's Fashion Fun __________________ yesterday.

Name ______________________________ Date ______________

Clever Ways to Create

Use context clues to help you define the italicized words in the sentences below. Circle the correct definitions.

1. Terry *fabricated* some wild story about why his homework wasn't done.
   A. wrote down
   B. performed on stage
   C. made up
   D. listened to

2. I need to *compose* an original poem for English class.
   A. read aloud
   B. talk about
   C. research
   D. write

3. My dad's contracting company *constructs* houses and some commercial buildings.
   A. supervises
   B. lives in
   C. builds
   D. designs

4. We will have to *formulate* a plan for how to get you to town by noon.
   A. create
   B. agree on
   C. purchase
   D. select from published choices

5. Brainstorming will help the class *generate* ideas before we write.
   A. locate in text
   B. develop the best of
   C. come up with
   D. organize old items

6. I did not *initiate* the talks, but I'm willing to participate in dialogue.
   A. like
   B. agree with
   C. think of as wise
   D. start

7. The idea of the lightbulb *originated* with Thomas Edison.
   A. surprised
   B. began
   C. frustrated
   D. seemed silly to

8. In order to *execute* our treehouse building plan, we will need to gather rope, wood, hammers, and nails.
   A. record
   B. explain
   C. put into action
   D. get approval for

9. Stanley *perpetrated* the practical joke, but everyone in the class laughed.
   A. liked the best
   B. was responsible for
   C. explained
   D. witnessed

10. Let's *institute* your new speech-giving ideas into student council elections policy.
   A. do just one time
   B. get rid of
   C. appreciate
   D. establish as rule

Name ______________________________ Date ______________

The Power of Literacy

Use context clues to help you define the italicized words from the passage. Circle the correct definitions.

Frederick Douglass was an intelligent and *principled* man whose *literacy* and values defined his life. Born into slavery in about 1818, the expectation was that Douglass would perform *menial* labor his whole life. But that's not what happened.

His first twist of fate occurred when the wife of his master *illegally* taught him the alphabet. From there, Douglass discovered for himself the magic of the written word. He watched white children read and studied the writing of sailors under whom he worked. The next fortunate event occurred when a free black man lent him freedom papers and a sailor uniform. Douglass used the disguise and papers to escape to New York, where slavery was *prohibited.*

Literacy then began to play a large role in Douglass's life. He wrote an *autobiography* to draw attention to the conditions of slavery. The book was so *eloquently* written that some people questioned the *authenticity* of its author. How could a former slave write so well? Write well he did. His book became a best seller. The newspapers he edited after that were well received too.

Next, Douglass began to give speeches. He became famous for his oratory skills and his reform stances. He spoke against slavery and in favor of women's rights, including their right to vote. By the time of his death in 1895, he had written books and newspapers, spoken in many countries, and counseled two presidents on minority issues. Douglass's literacy served him—and his country—well.

1. principled (based on a sense of…)
   A. financial concerns
   B. rightness and fairness
   C. hard work
   D. fame

2. literacy
   A. morals
   B. ability to work hard
   C. ability to read and write
   D. determination

3. menial
   A. professional
   B. farm-related
   C. a large amount
   D. lowly and common

4. illegally
   A. against common practice
   B. happily
   C. reluctantly
   D. against the law

5. prohibited
   A. allowed
   B. not allowed
   C. discouraged
   D. encouraged

6. autobiography
   A. life story of self
   B. life story of famous person
   C. novel
   D. poetry collection

7. eloquently
   A. in the language of the wealthy
   B. poorly
   C. beautifully and effectively
   D. childishly

8. authenticity
   A. race
   B. gender
   C. genuine nature
   D. age

Name ______________________________ Date ______________

Mallowcrunch Bars

A Tom Sawyer Retell

Use context clues to help you define the italicized words from the story. Circle the correct definitions.

Jake thought it was cool when his parents bought the corner *convenience* store in their little town of Rivercreek. It sold everything a person could want. Best of all, the store sold Jake's favorite candy—Mallowcrunch bars. What Jake hadn't realized was that when something was called a family business, it meant the whole family worked in it. His mom opened the store mornings, and his dad had it afternoons and evenings. Jake put in three hours after school and full days on weekends. He even had to pay for his own candy!

The first week of summer was especially *hectic*. It was busy all day! Jake hoped that at least he would be able to go fishing on Saturday.

"Sorry, kid," said his father on Saturday morning. "The grocery truck is unloading right now. All those cases have to be worked. And the returned soda cans and bottles have to be sorted. Drivers will be picking them up on Monday."

When Jake's three friends showed up, Jake was stocking cans of tomato soup.

"Where's your fishing pole?" asked one of them. "We're headed for the river bridge!"

Jake *concentrated* on getting the soup cans perfectly straight, their labels lined up. After a while he looked up. "You guys go ahead," Jake said. "I'd rather do this."

"You'd rather work than fish?" his friends gasped.

"You don't understand," said Jake. "I begged my dad to let me do this. It's a lot more fun than you think. After this, Mom's gonna let me sort the returned soda cans by vendor!"

Jake's friends stared at him, then each other. "You're joking, right?" said one of them. Jake shook his head. "Heck," the friend said, "let me try doing it, then."

"Nope," said Jake. They all argued with him, but Jake kept shaking his head. Finally he *relented*. "All right. For a Mallowcrunch bar each, you can stock for a little while."

That afternoon, and every Saturday afterward, customers at the store noticed how fully and neatly the shelves were stocked. And if they should happen by the Rivercreek bridge, they might notice four young boys leaning on the bridge rail, fishing *contentedly*. A close look might even *reveal* that one boy's pockets were full of Mallowcrunch bars.

1. convenience
   - **A.** large
   - **B.** small
   - **C.** quick and easy
   - **D.** full and busy
2. hectic
   - **A.** slow
   - **B.** chaotic
   - **C.** easy
   - **D.** hot
3. concentrated
   - **A.** ignored
   - **B.** enjoyed
   - **C.** tried not to think about
   - **D.** focused on
4. relented
   - **A.** gave in
   - **B.** disagreed
   - **C.** sighed
   - **D.** yelled
5. contentedly
   - **A.** unsatisfied
   - **B.** satisfied
   - **C.** angrily
   - **D.** jealously
6. reveal
   - **A.** hear
   - **B.** hunger
   - **C.** happily
   - **D.** show

Name ______________________ Date ____________

Think About It

The sentences below contain italicized words related to thinking. Use context clues to help you decide if the statements below the sentences are true or false. Circle **T** for true or **F** for false.

T F 1 I will *ponder* your recommendation before making a decision.
➧ A synonym for *ponder* is consider.

T F 2 I *assumed* there were four brothers from the picture, but there are only three. The fourth man in the picture is a cousin.
➧ An *assumption* is an educated guess.

T F 3 Ted *contemplated* moving to Nebraska, but he changed his mind.
➧ A synonym for *contemplated* is *decided*.

T F 4 Grandma likes to *reminisce* about life before the computer and microwave.
➧ *Reminisce* means to talk about a future time.

T F 5 My teacher could not even *conceive* of the idea of not assigning homework.
➧ *Conceive* means grasp or imagine in this sentence.

T F 6 Somehow, Sue has to *digest* everything in the chapter before the test tomorrow.
➧ *Digesting* the chapter suggests Sue intends to eat pages out of her book.

T F 7 I put a new battery in our flashlight so it would *emit* enough light for us to take a night hike.
➧ *Emit* is a verb meaning *throw* in this sentence.

T F 8 Mom *scrutinized* every inch of the antique chair and found it to be authentic.
➧ It would be a waste of time to *scrutinize* an important test before turning it in.

T F 9 I completed my science project, but I couldn't *interpret* its results.
➧ In this sentence, *interpret* means "translate from one language to another."

T F 10 I examined my doctor's note, but I couldn't *decipher* his handwriting.
➧ *Decipher* is a synonym for *duplicate*.

T F 11 When I'm at the dentist, I try to *visualize* being at the beach.
➧ *Visualize* and *vision* are related words.

T F 12 I *presume* you will meet me at the museum at noon.
➧ The word *presume* is related to the word *assume*.

Name ______________________________ Date ______________

Fun in the Garden

Use context clues to help you place the gardening words from the box into the sentences below.

WORD BOX

harvest
enlarge
flourish
thresher
shoot
mature
nourish
emerge
convert
photosynthesis
wither
pollinate

1. If you give your new plant water, sunlight, and nutrients, it should ____________________.
2. I'll cut a healthy ____________________ off my plant so you can have one of your own.
3. I keep bee hives beside my garden, and the bees ____________________ my flowers.
4. My petunias began to ____________________ when we had that dry spell last month.
5. Before planting your garden, add fertilizer to ____________________ your soil so your plants will grow tall and strong.
6. After the first bud, it is only a few days before your blossoms will ____________________.
7. Plants ____________________ sunlight into energy.
8. The process is called ____________________.
9. Even after you see a pumpkin growing on the vine, it will take weeks for it to ____________________.
10. You can ____________________ your pumpkin by cutting a small hole and adding a little milk just as it starts to turn orange. By fair time, it will be huge.
11. If you plant a large garden and keep up with the weeding, you should get a great ____________________ in the end.
12. Can you imagine having to beat the husks of grains by hand before the invention of the mechanical ____________________?

Name ______________________________ Date ______________

Now That's Worth Something

Use context clues to help you define the italicized words about value. Circle the correct definitions.

1. Jenny kept the *priceless* diamond necklace her grandma gave her in a safe.
   A. not having a price
   B. free
   C. cherished and expensive
   D. worthless

2. One of Stan's most *admirable* qualities is his ability to give compliments.
   A. to be admired
   B. annoying
   C. talented
   D. frequent

3. Our *esteemed* great-grandfather always sits at the head of the table.
   A. ancient
   B. feared
   C. respected
   D. only

4. The quilt hanging on the wall is an *heirloom* from your great-aunt.
   A. blanket
   B. craft item
   C. inherited money
   D. inherited antique

5. Oil is a valuable *commodity* in our automobile-centered society.
   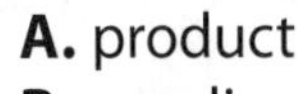
   A. product
   B. gasoline
   C. energy source
   D. concern

6. Your commitment to walk to work each day is *commendable*; I'm impressed.
   A. ridiculous
   B. pertaining to fitness
   C. time-consuming
   D. to be praised

7. The Joneses have been *prominent* in our town since their ancestors founded it.
   A. absent
   B. widely known
   C. disliked
   D. bossy

8. Is this a *suitable* outfit for the interview?
   A. informal
   B. formal
   C. colorful
   D. appropriate

9. It is of *paramount* importance that you get this to the bank by five o'clock.
   A. little
   B. no; none
   C. ultimate
   D. some

10. Your *exquisite* necklace will be beautiful with your wedding gown.
    A. diamond
    B. pearl
    C. inexpensive
    D. exceptionally beautiful

Name ______________________________ Date ______________

Historically Speaking

Use context clues to help you define the italicized history words in the sentences below.

1. ______ The *collapse* of the Soviet Union resulted in the naming of many new countries.
2. ______ It also improved relations between the Soviets and the Americans, ending the *Cold War.*
3. ______ The *emancipation* of slaves was one of the conflicts at issue during the Civil War.
4. ______ The Internet helps *interstate* business owners do business with less travel.
5. ______ After Pearl Harbor, Japanese Americans were forced to live in *internment* camps.
6. ______ The Congress and the House of Representatives make up the *legislative* branch of the U.S. government.
7. ______ To change anything in the U.S. Constitution, the people need to pass an *amendment.*
8. ______ Blacks were not allowed to hold government jobs during *apartheid* in South Africa.
9. ______ Blacks were not allowed to eat in some restaurants during *segregation* in the U.S.
10. ______ *Refugees* from Sudan have been living in tents in neighboring countries for decades.
11. ______ The U.S. president can grant *amnesty* to anyone, but it is usually reserved for political offenses.
12. ______ Adults in a democracy have a *civic* duty to vote.

DEFINITIONS

A. people who live away from their homeland due to war or natural disaster

B. sudden failure of a system

C. confinement during war

D. involving more than one state

E. legal separation of ethnic groups in U.S. history

F. a change made to the Constitution

G. unfriendly relations between nations

H. freedom

I. having to do with laws

J. concerning citizens

K. legal separation of ethnic groups in South Africa's history

L. an official pardoning of a crime

Name ______________________________ Date ______________

Back to Nature

Use context clues, including your understanding of sentence structure, to help you choose the correct words from the boxes to fill in the blanks in the passages below about nature.

A Because the ________ (1) wants to study birds and fish in their natural ________ (2), he works to ________ (3) their ________ (4) on Highway 1. His daughter, who likes the comfort of a controlled temperature and indoor environment, would rather visit birds in the ________ (5).

WORD BOX
habitat
refuge
aviary
preserve
naturalist

B The Klamath ________ (6), in south-central Oregon, is a large, bowl-shaped, low-lying stretch of land. It is home to some towns, but most of the region consists of ________ (7) communities that depend on agriculture for their livelihood. One of the most important natural ________ (8) of the region is its forests. Besides providing lumber, the forests are also used for camping, fishing, and the hunting of deer and various species of ________ (9).

WORD BOX
fowl
Basin
resources
rural

C I live in a big city with an extensive park system where I can enjoy nature. I love to rest under the trees at the ________ (10) and to watch the birds fly freely in their protected ________ (11). The protected bird zone is maintained by a local ________ (12) who studies birds as both a hobby and an occupation. In another section of the park system, I observe water plants and animals at a ________ (13) that is protected by law from being drained to provide more building space.

WORD BOX
marsh
arbor
sanctuary
ornithologist

Name ______________________________ Date ______________

Which Way Did You Say?

Use context clues to help you locate the best synonym for each italicized word in the sentences below.

1. ______ The train *timetable* made it easy for Erin to plan her day trips in England.
2. ______ Henry plans to *immigrate* to France in the next few years.
3. ______ You will need your *passport* to get into Mexico.
4. ______ The travel *agency* told me to book my trip early.
5. ______ Rick's *nautical* ship was built for the ocean, not this little river.
6. ______ Liv is an *expatriate* from Denmark.
7. ______ Arapahos are *indigenous* to the United States; Europeans traveled to get to America.
8. ______ We had to pay a one dollar and fifty cent *toll* before we could get on the bridge.
9. ______ The water is so shallow that you can *traverse* the creek on foot.
10. ______ My daily *commute* to work takes about one hour each way.
11. ______ I'm preparing for a four-day biking *excursion* that will take me across the state.
12. ______ I would rather go *abroad* by plane than by ship.

SYNONYMS

A. cross
B. schedule
C. identification
D. fee
E. emigrant
F. seagoing
G. travel
H. overseas
I. move
J. company
K. journey
L. native

Name ____________________ Date ____________

Check Out the News

Use context clues, including your understanding of sentence structure, to help you complete the newspaper-related passages with words from the boxes.

A I ran an ad in the ____________ (1) trying to sell my old car. The ad ____________ (2) on Saturday and ran for one week. Although it seemed like a great ____________ (3) for someone to get a good car cheaply, no one called about my ad. I guess I should have first questioned the ____________ (4) ____________ (5) about how best to word my ad.

WORD BOX
advice
classifieds
opportunity
debuted
columnist

B Cindy wrote a letter of complaint to the ____________ (6) of the newspaper after her ____________ (7) price rose by 20%. Since she pays for ____________ (8) of the paper to her door, she expects to pay a fair price. Still, as a regular customer, she expects a ____________ (9) off the ____________ (10) price, and with the new price hike, she is no longer getting it.

WORD BOX
delivery
newsstand
editor
discount
subscription

C The ____________ (11) word on the winner of this year's Academy Awards is not yet out, but every reporter at the newspaper has an opinion. The movie review page is running ____________ (12) from every movie reviewer on the planet about who might win. The ____________ (13) news page talks about local hero Jake Blade's appearance at the premiere of his ____________ (14) movie. The ____________ (15) mention that the late Suzette Nelson performed in a contender for the award before her recent death. The ____________ (16) page, which usually runs pieces about local celebrations only, even covered the upcoming award ceremony in a short article comparing it to a local wedding. And, of course, both the comic strips and the ____________ (17) are sharpening their satire about the time-honored tradition of watching the Academy Awards.

WORD BOX
society
obituaries
commentary
regional
official
nominated
humorists

Answer Key

Word Part Clues (page 4)

1. A
2. C
3. B
4. D
5. B
6. B
7. B
8. A
9. C
10. B

Example Clues (page 5)

1. Alternative
2. controversial
3. Primitive
4. ambitious
5. grievances
6. Vendors
7. collaborative
8. audition
9. diverse
10. feats
11. Opposition
12. endurance

Definition and Synonym Clues (page 6)

1. old objects
2. competitor
3. old
4. deceit or dishonesty
5. treating a group of people unfairly
6. the right to vote
7. well-crafted copy
8. salesperson
9. convince
10. gathering of people
11. freedoms
12. committed; required
13. strict
14. break

Antonym and Contrast Clues (page 7)

1. C
2. A
3. C
4. B
5. B
6. C
7. A
8. B
9. A
10. C

Sentence Structure Clues (page 8)

1. A
2. B
3. C
4. B
5. C
6. A
7. A
8. B
9. B
10. A

Background Knowledge Clues (page 9)

1. C
2. A
3. C
4. A
5. B
6. B
7. D
8. A

Context Clues Collection (page 10)

1. example clue: A
2. definition/synonym clue or sentence structure clue: C
3. background knowledge clue: B
4. antonym/contrast clue: C
5. word part clue: A
6. sentence structure or background knowledge clue: C
7. definition/synonym clue or sentence structure clue: C
8. example clue: D

The Medical World (page 11)

1. B
2. K
3. C
4. F
5. A
6. G
7. H
8. D
9. I
10. E
11. J
12. L

Grasping Government (page 12)

1. C
2. A
3. B
4. B
5. A
6. C
7. A
8. D
9. D
10. C

Interpreting Action (page 13)

1. B
2. I
3. L
4. K
5. C
6. A
7. H
8. F
9. E
10. J
11. G
12. D

Plenty of Personality (page 14)

1. tiresome
2. explosive
3. perfectionist
4. insistent
5. distress
6. reserved
7. treacherous
8. reliable
9. aggression
10. husky
11. lavish
12. magnetic

Visual Vocabulary (page 15)

1. true
2. false
3. true
4. true
5. true
6. false
7. false
8. true
9. false
10. true
11. true
12. true

Board Game Bonanza (page 16)

1. A
2. D
3. D
4. B
5. A
6. B
7. C
8. A
9. B
10. B
11. C

What's with the Weather? (page 17)

1. frigid
2. snowshoeing
3. inclement
4. recreation
5. avalanche
6. sweltering
7. monsoon
8. meteorologist
9. typhoon
10. blazingly
11. sultry
12. refreshed
13. slathering

Community Service (page 18)

1.	B	5.	D
2.	C	6.	C
3.	A	7.	A
4.	C	8.	C

Space-tacular Words (page 19)

1.	C	6.	A
2.	B	7.	B
3.	A	8.	B
4.	C	9.	C
5.	D	10.	D

The Generals Who Became Emperors (page 20)

1.	A	7.	C
2.	B	8.	D
3.	A	9.	C
4.	B	10.	D
5.	D	11.	B
6.	C	12.	A

How Do They Do That? (page 21)

1. literally
2. heartily
3. religiously
4. vigorously
5. frantically
6. mildly
7. seldom
8. fanatically
9. solemnly
10. gravely
11. scholastically
12. frankly

Water Words (page 22)

1.	K	7.	A
2.	B	8.	H
3.	E	9.	D
4.	G	10.	J
5.	C	11.	L
6.	F	12.	I

Preferring Prefixes (page 23)

1.	C	6.	B
2.	B	7.	B
3.	A	8.	C
4.	D	9.	D
5.	A	10.	C

Mr. Fribble Learns Too Late (page 24)

1.	B	5.	A
2.	A	6.	C
3.	C	7.	D
4.	C	8.	D

A Job Well Done—or Otherwise (page 25)

1.	G	7.	L
2.	E	8.	J
3.	F	9.	I
4.	A	10.	B
5.	H	11.	C
6.	K	12.	D

A Display of Attitude (page 26)

1. domineering
2. confident
3. collective
4. prejudice
5. congenial
6. insecure
7. submissive
8. opinionated
9. jovial
10. arrogant
11. assertiveness
12. manipulated

Talk About Time (page 27)

1.	B	6.	C
2.	A	7.	C
3.	D	8.	C
4.	B	9.	D
5.	B	10.	C

The Ant (page 28)

1.	C	6.	B
2.	B	7.	A
3.	A	8.	B
4.	D	9.	B
5.	A	10.	A

Eatin' Up Vocabulary (page 29)

1.	B	7.	C
2.	J	8.	L
3.	G	9.	K
4.	F	10.	A
5.	I	11.	E
6.	H	12.	D

Just Playin' Around (page 30)

1. swerved
2. collisions
3. distinguished
4. dismay
5. scheme
6. regulations
7. standard
8. equipped
9. darted
10. participate
11. division
12. astounding

Adventure in New Lands (page 31)

1.	A	7.	A
2.	B	8.	A
3.	D	9.	B
4.	C	10.	C
5.	D	11.	A
6.	A	12.	D

Now That's Entertainment (page 32)

1.	C	6.	C
2.	C	7.	D
3.	C	8.	A
4.	A	9.	B
5.	A	10.	C

Challenges of Nature (page 33)

1.	B	**7.**	B
2.	C	**8.**	D
3.	D	**9.**	B
4.	A	**10.**	D
5.	D	**11.**	A
6.	D		

Dozens of Disciplines (page 34)

1.	true	**7.**	true
2.	false	**8.**	true
3.	true	**9.**	false
4.	false	**10.**	true
5.	true	**11.**	true
6.	false	**12.**	true

Troubleshooting Technology (page 35)

1. font
2. graphics
3. hacker
4. troubleshoot
5. cyberspace
6. banners
7. home page
8. search engine
9. formatting
10. insert
11. document
12. blog

Clever Ways to Create (page 36)

1.	C	**6.**	D
2.	D	**7.**	B
3.	C	**8.**	C
4.	A	**9.**	B
5.	C	**10.**	D

The Power of Literacy (page 37)

1.	B	**5.**	B
2.	C	**6.**	A
3.	D	**7.**	C
4.	D	**8.**	C

Mallowcrunch Bars (page 38)

1.	C	**4.**	A
2.	B	**5.**	B
3.	D	**6.**	D

Think About It (page 39)

1.	true	**7.**	true
2.	true	**8.**	false
3.	false	**9.**	false
4.	false	**10.**	false
5.	true	**11.**	true
6.	false	**12.**	true

Fun in the Garden (page 40)

1. flourish
2. shoot
3. pollinate
4. wither
5. nourish
6. emerge
7. convert
8. photosynthesis
9. mature
10. enlarge
11. harvest
12. thresher

Now That's Worth Something (page 41)

1.	C	**6.**	D
2.	A	**7.**	B
3.	C	**8.**	D
4.	D	**9.**	C
5.	A	**10.**	D

Historically Speaking (page 42)

1.	B	**7.**	F
2.	G	**8.**	K
3.	H	**9.**	E
4.	D	**10.**	A
5.	C	**11.**	L
6.	I	**12.**	J

Back to Nature (page 43)

1. naturalist
2. habitat
3. preserve
4. refuge
5. aviary
6. Basin
7. rural
8. resources
9. fowl
10. arbor
11. sanctuary
12. ornithologist
13. marsh

Which Way Did You Say? (page 44)

1.	B	**7.**	L
2.	I	**8.**	D
3.	C	**9.**	A
4.	J	**10.**	G
5.	F	**11.**	K
6.	E	**12.**	H

Check Out the News (page 45)

1. classifieds
2. debuted
3. opportunity
4. advice
5. columnist
6. editor
7. subscription
8. delivery
9. discount
10. newsstand
11. official
12. commentary
13. regional
14. nominated
15. obituaries
16. society
17. humorists